Welcome to our *HappyStoryGarden!*

Copyright © 2024 by Viktoriia Harwood

All rights reserved.

No part of this book may be reproduced in any form or by any electronic or mechanical means, including information storage and retrieval systems, without written permission from the author, except for the use of brief quotations in a book review.

2024

Victoria Harwood

WHO WILL I BE WHEN I GROW UP?

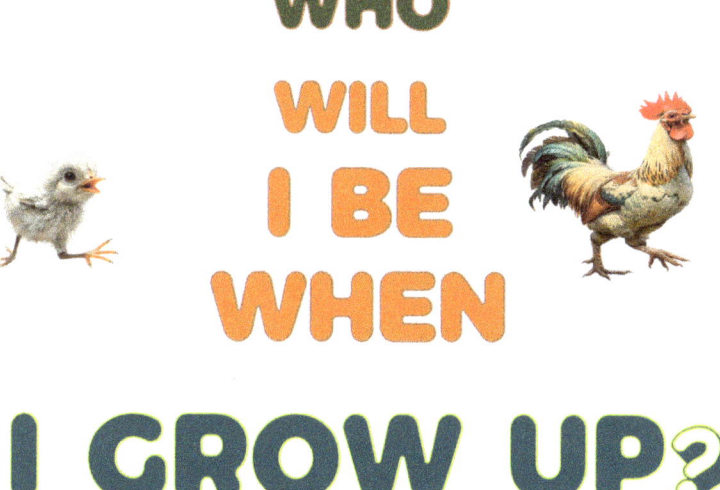

All animals in our world live in families.
All youngsters grow up and become as magnificent, smart and elegant as their parents.

For example, a foal will grow into a beautiful horse, and a gosling will grow into a long-necked, proud goose; he will also fly to distant lands, like his parents.

Look at each baby and find out what they will become when grown up.

Every baby from the animal kingdom must learn to live in the world or survive in the wild; their life can be full of dangers and adventures.
Now, you and I are going on a journey called "Who will I be when I grow up?"

You will meet 40 animals and their little ones:

Hare	Hedgehog
Goose	Cow
Hamster	Walrus
Frog	Tortoise
Dolphin	Chimpanzee
Hippopotamus	Kangaroo
Whale	Rhinoceros
Lion	Bison
Goat	Squirrel
Wolf	Mouse
Crocodile	Elephant
Bear	Turkey
Deer	Sheep
Tiger	Monkey
Beaver	Horse
Cat	Pig
Dog	Seal
Guinea Pig	Duck
Camel	Chicken
Fox	
Giraffe	

They call me a Chick, and I know that I am a bird.
When I grow up, I will have wings and a red comb on my head, I will live in a shed and lay eggs.
What will **I grow up to be?**

Chicken

I am a bird and they call me a duckling.
When I grow up, I will be able to fly, swim and dive and eat algae from the bottom of a pond.
What will I grow up to be?

Duck

They call me a Pup but I'm not a dog.
When I grow up, I will be big and fat, swim very well and fish underwater. I will lie on the shore of a the sea in a group.
What will I grow up to be?

Seal

They call me a piglet.
When I grow up, I will be big and handsome with short bristly hair and big ears. I will learn to dig the ground with my snout in search of tasty roots.
What will they call me?

Pig

They call me a Foal now and I can already run fast and love fresh grass and carrots.

When I grow up, I will help people and give them rides on my back.

What will I grow up to be?

Horse

They call me an Infant.
When I'm older, I will live in a group and will jump from one tree branch to another with ease and look for delicious fruits to eat.
What will I grow up to be?

Monkey

They call me a Lamb or Lambkin.

When I grow up, I will live in a field and give people a lot of wool.

What will I grow up to be?

Sheep

They call me a Leveret and I live in a warm nest hidden in tall grass.

When I become an adult, I will grow long ears and strong hind legs, and I will learn to run quickly through fields and forests.

Who will I be?

Hare

They call me a Poult.
When I'm older I will become a large bird with dark plumage and live on a farm.
What will I grow up to be?

Turkey

They call me a Calf.
When I grow up, I will be big, kind, strong, and have a long trunk and two tusks.
What will I grow up to be?

Elephant

They call me a Pinkie and I live in a nest with other little ones just like me in a field.

When I grow up, I will dig myself a hole and stock up on tasty grains.

What will I grow up to be?

Mouse

They call me a Kit and I live in a hollow high up in a tree. When I grow up, I will climb trees easily, collect nuts and mushrooms and hide them for the winter.

What will I be when I grow up?

Squirrel

They call me a Calf.
When I grow up, I will become a large shaggy-haired animal and live in a large group.
Who will I be?

Bison

They call me Calf now but when I grow up, I will be big and strong with tough thick skin and have a horn.
What will I grow up to be?

Rhinoceros

They call me a Joey and now I live with my mother in a pouch on her stomach.

When I grow up, I will be strong and become expert at jumping far.

What will I grow up to be?

Kangaroo

They call me a Hatchling now.
I hatched from eggs that my mother buried in the ground.
My shell will become stronger and larger when growing up.
Who will I become?

Tortoise

They call me an Infant.
When I grow up, I will be intelligent, strong and agile.
What will I grow up to be?

Chimpanzee

They call me a Calf.
When I grow up, I will become big and fat and have long whiskers.
Who will I grow up to be?

Walrus

They call me a Calf.
I will live on a farm and give people milk when I am an adult.
What will I grow up to be?

Cow

They call me a Hoglet.
When I grow up, I will be an agile hunter at night and have prickly coat on my back.
What will I grow up to be?

Hedgehog

They call me a Calf and I live in the African continent.
When I grow up, I will be very tall and able to reach leaves at the top of trees.
What will be when I grow up?

Giraffe

They call me a Cub.
When I grow up, I will be an agile and cunning hunter, I will have a fluffy red tail.
What will I grow up to be?

Fox

They call me a Calf.
When I grow up, I will be smart and strong and help people carry things a long distance in the desert.
What will I grow up to be?

Camel

They call me a Pup.
When I grow up, I will store seeds in my cheeks; I will be fat and good-natured.
Who will I be?

Hamster

They call me a Pup.
When I grow up, I will have silky fur and long front teeth, and everyone will love me.
Who will I be?

Guinea Pig

They call me a Puppy.
When I grow up, I will be bright, bark loudly and help people guard the house.
What will I grow up to be?

Dog

They call me a Kitten.
When I grow up, I will be intelligent, bright and playful.
What will I grow up to be?

Cat

They call me a Kit.
When I grow up, I will learn to cut trees to build dams on rivers.
What will I grow up to be?

Beaver

They call me a Cub.
When I grow up, I will be strong and fierce.
What will I become?

Tiger

They call me a Fawn.
When I'm older I will look like the king of the forest with my tree-like horns.
Who will I be?

Deer

They call me a Cub.

Now I am small, but when I become an adult, I will be huge, shaggy and fearsome. I will learn to climb trees and fish in rivers. I also love honey.

What will I be when I grow up?

Bear

They call me a Hatchling.
I will be the biggest and strongest animal in the river when I grow up.
What will I grow up to be?

Crocodile

They call me a gosling, and I am a bird.
When I grow up, I will have big wings and a long neck and I will fly high in the sky, and for the winter my whole flock will migrate to warmer places.
Who will I be?

Goose

They call me a Pup and I live in the forest.
When I grow up, I will be a hunter and live in a pack.
All animals will fear and respect me.
What will I grow up to be?

Wolf

They call me a Kid.
When I grow up, I will become beautiful, eat a lot of grass, and give people milk like my mother.
Who will I be?

Goat

They call me a Cub.
When I grow up, I will be a strong and formidable hunter and be known as the King of Beasts
What will I become?

Lion

They call me a Calf.
When I grow up, I will be magnificent and huge; I will swim in the oceans and eat plankton.
What will I grow up to be?

Whale

They call me a Calf.
When I grow up, I'll be huge and strong with a thick skin and like bathing in mud.
What will I grow up to be?

Hippopotamus

They call me a Calf.
My mother feeds me milk. When grown up, I will become a fast swimmer in the seas and oceans and be a cheerful creature.
What will I grow up to be?

Dolphin

They call me a Tadpole and I live in a pond.
When I'm an adult, I will catch flies and mosquitoes with my long tongue, jump high.
What will I grow up to be?

Frog

Thank you for taking this exciting journey to discover so many animals and their babies with me!

Well done!

I hope this book brought you a lot of joy and new knowledge. The natural world is rich and amazing, and there is always something new to learn.

We humans must protect nature and take care of animals, birds and other living creatures.

Special thanks

Special thanks to Igor Kirko and Zinaida Kirko
for the practical advice and support in
creating the book and illustrations.

My thanks also to Leslie Harwood, an
excellent translator and kind editor.

WELCOME TO
THE HAPPY STORY GARDEN

www.ingramcontent.com/pod-product-compliance
Lightning Source LLC
LaVergne TN
LVHW070205080526
838202LV00063B/6564